Prospects for Early Deployment of Power Plants Employing Carbon Capture

John Ruether[1], Robert Dahowski[2],
Massood Ramezan[3], and Charles Schmidt[1]

1. National Energy Technology Laboratory, USDOE
2. Pacific Northwest National Laboratory, Battelle
3. National Energy Technology Laboratory, SAIC

SUMMARY

Coal-based IGCC with CO_2 capture and sequestration would yield only one fifth the specific carbon emissions (kg C or kg CO_2 /kWh) as would state-of-art NGCC. California appears to be a good venue for consideration of IGCC+S: there is need for additional generating capacity and an unserved market for CO_2 that could be used to conduct enhanced oil recovery. In this paper, a probabilistic analysis is conducted to determine Required Selling Price of Electricity (RSPOE) and expected rate of return on common stock equity for three fossil generating technologies: NGCC, NGCC+S (NGCC with capture and sequestration), and IGCC+S. Variables treated probabilistically are the costs of natural gas and coal fuels, and the values of electricity and CO_2 products. Predictions of prices prepared by the Energy Information Agency are used together with measures of price variability based on historic price fluctuations. Installation of new generating plant is assumed to occur in 2010 and operate for a 20 year book life to 2030.

It is shown that when CO_2 can be sold at historically realized prices for use in enhanced oil recovery (EOR), IGCC+S is expected to be profitable with no subsidy for avoidance of CO_2 emissions. Expected profitability of NGCC is greater than that of IGCC+S, but so is the uncertainty of RSPOE and expected rate of return on common stock equity, due principally to uncertainty of natural gas price. NGCC+S exhibits both a higher RSPOE and higher uncertainty of RSPOE than either of the other technologies.

INTRODUCTION

Stabilization of atmospheric concentration of greenhouse gases, of which CO_2 is the most important, "….at a level that would prevent dangerous anthropogenic interference with the climate system…"[1] is a widely accepted policy goal. When concerted actions start to be taken to achieve this goal, fossil generating stations, as large point sources of CO_2, may be required to make disproportionately large emission reductions because doing so will be cost effective. At present natural gas combined cycle (NGCC) is the technology of choice for providing new electric generating capacity in the U.S. for reasons that include environmental performance, thermal efficiency, high availability compared to

---

[1] From Article 2 of the UN Framework Convention on Climate Change (Rio Accord).

renewables, and relatively low capital cost. Relatively low specific carbon emissions (kg C or kg CO_2/kWh) compared to coal generators is another attraction of NGCC. Yet NGCC cannot be the only response of the electric power industry to the challenge of global warming even if affordable supplies of natural gas were assured into the indefinite future. Climate modelers estimate that upwards of 60% reduction in greenhouse gas emissions *from current levels* will be needed to stabilize atmospheric composition. That is a greater reduction than could be achieved even if all coal –fired units were replaced with state-of-art NGCC.

This paper invites serious consideration of fossil fueled electricity generation technologies that capture nominally 90% of CO_2 emissions and use the CO_2 to conduct enhanced oil recovery. Carbon sequestration of this kind represents a fundamentally different approach to reducing carbon emissions that has potential not less than traditional approaches such as improvement of thermal efficiency of generation, improvement of end use efficiency, and use of renewables. There is no immediate prospect for commercial deployment of fossil generation with CO_2 capture and sequestration, however, because with no value assigned to reducing carbon emissions, such processes are more expensive than conventional fossil generation. One approach to overcoming this problem is to investigate use of a carbon tax or carbon emission cap. This study takes a different approach. It considers how the economics of natural gas- and coal-based generation with carbon capture would fare if a market for the collected CO_2 is assured for practice of EOR.

In the following, the practice of EOR in the U.S. is reviewed, and attention is paid to prospects for providing both needed new generating capacity and CO_2 for the practice of EOR in California. Then the results of a financial analysis of natural gas-based and coal-based generating technologies with and without carbon capture for potential installation in California starting in 2010 are presented.

CO_2 EOR: OVERVIEW AND PROSPECTS

Carbon dioxide enhanced oil recovery (CO_2 EOR) is one of several methods to enhance the production of oil from mature reservoirs whose output is declining under normal production processes. It has been the fastest growing EOR method, and currently accounts for about 25% of U.S. total EOR production. The most common CO_2 EOR method is miscible displacement, in which the injected CO_2 dissolves in the oil, increasing its volume and reducing its viscosity. This increases the mobility of the oil, resulting in the production of oil bypassed by primary and secondary recovery methods. Typical CO_2 floods, under the right conditions, can yield an additional 7 to 15 percent of original oil in place (OOIP), extending the life of a producing field by as much as 15-30 years (Moritis, 2001).

The United States is the world leader in the development and application of CO_2 EOR. In fact, commercial practice began in West Texas in 1972, and continues to flourish there today. According to a 2000 EOR survey, there are a total of 64 CO_2 projects in the U.S., 47 of these in the Permian Basin area of West Texas and Southeast New Mexico. Other areas with activity include the Rocky Mountain region, Oklahoma, and Mississippi. Collectively, these projects produce over 190,000 barrels of incremental oil per day (bbl/d), accounting for 3.3% of total U.S. crude production. Two additional projects have recently come online and several more are being planned. Slide 2 shows the locations of active CO_2 EOR projects (in purple) along with several planned and pilot sites (red oil derricks).

The CO_2 used at these fields comes from several different sources. Most is supplied by large underground deposits of naturally occurring and high purity CO_2 (shown as the large green spots on Slide 2). Three such domes presently serve the fields of the Permian Basin with over 1 billion cubic feet per day (Bcf/d) of 97-99% pure CO_2, and have recoverable reserves estimated at over 12 trillion cubic feet (Tcf). This CO_2 is delivered to the fields via an extensive network of dedicated pipelines. A smaller number of projects utilize CO_2 waste streams from industrial sources including natural gas processing facilities and fertilizer plants.

Prospects for growth and expansion of CO_2 EOR look promising. Analyst estimates for the Permian Basin indicate that over 50 additional projects, adding 500 million to 1 billion barrels of oil reserves, are economically viable at recent prices and current technology. One operator in the Permian Basin is planning to initiate 4-5 new projects in the next five years, in addition to 10-12 expansions of existing projects (ibid.). Others likely have similar plans.

Several other key areas are believed to be ripe for CO_2 injection as well, but have to this point lacked a dependable supply of inexpensive CO_2. Where natural sources are not available, operators have been reluctant to gamble on a CO_2 flood. However, several projects are underway that could lead to a vast expansion of this EOR technology. A pipeline carrying waste CO_2 from the LaBarge natural gas plant in Wyoming is being extended further towards numerous fields in Central and Northern Wyoming (ibid.). In Central Kansas, a field demonstration sponsored in part by the U.S. Department of Energy (DOE) will examine the technical and economic feasibility of CO_2 flooding to recover residual oil from mature reservoirs in that region (www.kgs.ukans.edu). This will be the first time CO_2 has been used for EOR in Kansas, and if successful, could lead to the development of CO_2 supplies and the possible additional recovery of over 250 million barrels of incremental oil.

Yet, even with all of the action in Wyoming and Kansas, many industry experts believe that the next largest opportunity for CO_2 flooding beyond the Permian Basin exists in California. The fourth largest oil-producing state in the U.S., California has many large mature fields that may respond well to CO_2 injection; one recent estimate of demand was on the order of 3-5 Tcf of CO_2. While no large, stable supply of CO_2 is readily available, operators in the San Joaquin basin are considering this EOR technique to boost production. In another DOE-sponsored project, Chevron is in the midst of conducting a pilot injection study at their Lost Hills field. The field, discovered in 1910, has had a cumulative oil production of only 135 million barrels or 5% of OOIP, largely due to its low permeability. Under CO_2 injection, a rapid oil response has been observed and it is hoped that oil recovery can be increased to 20% of OOIP, effectively tripling overall production. If proven successful in this field, this technique could help recover billions of barrels of oil trapped in the siliceous shales and diatomite reservoirs of this rich petroleum province (Montgomery et al., 2000).

CO_2 for this California pilot project is being trucked over 120 miles to the injection site at a cost of $3.50/Mcf. This illustrates both the importance of the project to the oil resource base of this region as well as the need to secure a convenient CO_2 supply. In order to meet this anticipated need for CO_2, Ridgeway Petroleum is considering building a pipeline from its newly discovered deposits of highly concentrated CO_2 (plus helium) beneath the Arizona/New Mexico border region. The St. John's formation contains an estimated 14.8 Tcf of CO_2 in place, along with 64 Bcf of helium (http://www.ridgewaypetroleum.com/news/arizona.html). However, the pipeline would need to be some 600 miles in length and cross some very mountainous terrain, making it a costly and potentially risky endeavor. Ridgeway Petroleum are therefore evaluating the potential California CO_2 market.

The economics of a CO_2 EOR project is heavily tied to the price of oil and availability of CO_2. CO_2 purchases constitute the single largest cost of a CO_2 EOR project (even at the lowest cost of natural CO_2 at about $0.65/Mcf). A reliable, nearby source of CO_2 is a key for oil field operators to consider CO_2 injection. Production response and effectiveness of enhancement is highly reservoir specific with net utilization rates typically in the range of 2.5 – 11 Mcf CO_2 injected per bbl incremental oil produced, averaging about 6 Mcf/bbl. At present, the costs of delivering CO_2 from various sources are roughly as follows: $0.65/Mcf from natural domes, $1/Mcf from natural gas processing, and $3/Mcf when captured from power plant flue gas. These values represent floor prices for providing CO_2 from the respective resources.

While the cost of capturing CO_2 from power plant flue gas is still considered by most to be too expensive for EOR, in places where cheaper sources of CO_2 are not available it can make economic sense. A study conducted by the Bureau of Economic Geology (BEG) in Texas (1999) examined the potential for capturing CO_2 from a group of 37 coal- and lignite-fired power plants and using it for EOR, in areas of Texas far from the existing CO_2 delivery infrastructure. The analysis identified over 1700 significant reservoirs in the state that could be used to sequester power plant CO_2 while enhancing oil production. A total of 3 billion barrels of residual oil was determined to be recoverable through CO_2 injection within only 30 miles of candidate power plants, and increasing to 6 billion barrels within 60 miles, and 8 billion within 90 miles. They concluded that substantial potential exists for CO_2 EOR in the mature oil reservoirs of Central and Eastern Texas using power plant emissions, and that such an approach may be cost-effective when compared to converting the plants to burn natural gas in order to reduce emissions. The BEG study considered retrofitting CO_2 capture equipment on pulverized coal generators. By contrast, in the present work we treat new generating capacity instead of retrofits, and IGCC instead of PC coal fired generators. The cost of CO_2 capture would be substantially lower for the approach considered in the present study.

The next few years will likely see strong growth in CO_2 EOR. It has been estimated that if pure and inexpensive CO_2 were available to all U.S. oil fields, total demand would be on the order of $60 - >100$ Tcf (Martin and Taber, 1992). Due to the disperse locations of the target fields and increasing urgency of reducing greenhouse gas emissions, utility plant CO_2 emissions may well become a growing part of the supply mix.

RESULTS

A probabilistic analysis was performed to determine the Required Selling Price for Electricity (RSPOE) for the period 2010-2030 for three technologies for electricity generation:

Natural Gas Combined Cycle (NGCC)
Natural Gas Combined Cycle with CO_2 capture and sequestration (NGCC+S)
Integrated Gasification Combined Cycle with CO_2 capture and sequestration (IGCC+S)

A probabilistic analysis was also performed for the expected rate of return on common stock equity. For both analyses, equations were developed that employ price predictions that contain uncertainty. MonteCarlo simulation was used to estimate expected values of RSPOE and expected rate of return on common stock equity, as well as standard deviations for these estimates.

All historic prices were converted to year 2000 dollars by use of values for the Gross Domestic Product Implicit Price Deflator. Furthermore, price predictions are also stated in year 2000 dollars. Thus all prices in this paper refer to year 2000 dollars, and computed values for RSPOE and rate of return on common stock equity are expressed in constant dollars.

We note that there are many kinds of risk that builders and operators of electricity generators face. Some of these are listed in Slide 5. In the present study we treat only Supply Risk and Market Risk. See Frey and Iwanski (1992) for a discussion of Technical Risk for IGCC power generation.

It was assumed for the two technologies that capture CO_2 that all CO_2 collected would be sold for use in enhanced oil recovery (EOR). Thus, NGCC+S and IGCC+S had two principal revenue streams, electricity and CO_2 , while NGCC had only one, electricity.

Data on both performance, e.g., heat rate or efficiency, and costs for all three generation technologies were obtained from a report prepared by Parsons Energy and Chemicals Group under sponsorship of EPRI and USDOE (Ref. *Evaluation of Innovative Fossil Fuel Power Plants with CO_2 Removal*, 2000). A summary of the characteristics of the three technologies modeled in this work, as well as IGCC without capture, is shown in Slide 6. The notation "H" in the names of the technologies refers to the use of that type of combustion turbine.

Calculations for Levelized Cost of Electricity (LCOE) are presented in the Parsons report for all three technologies modeled in the present work at two average capacity factors, 65% and 80%. The figures for LCOE shown in Slide 6 were computed by Parsons for natural gas cost and coal cost of $2.70 and $1.24, respectively, per million BTU (HHV). The figures for LCOE shown in Slide 6 do not include revenue for the sale of CO_2 . Costs in both the Parsons report and the present analysis are reported in year 2000 constant dollars.

Slide 6 shows that specific carbon emissions for power generated without capture via NGCC are less than half as large as those for IGCC without capture. With nominal 90% capture for both the gas- and coal-based technologies, specific carbon emissions for NGCC+S again is about half as large as IGCC+S. But notice also that specific carbon emissions for IGCC+S are only about one-fifth as large as for NGCC. Thus if reducing carbon emissions in the course of power generation becomes important, use of IGCC+S would represent a significant improvement compared to NGCC. Notice also that efficiency degradation is greater when capture is practiced with NGCC than with IGCC, and that the capital cost increment for providing capture is greater for NGCC than for IGCC. As explained in the Parsons report, both observations are due to the different manner in which capture is accomplished in the two generating approaches, from the flue gas with NGCC and from syngas with IGCC.

The size of plants for the three generating technologies studied here showed some variation. Net power output ranged from 310.8- 403.5 MW. Some discussion is presented in the Parsons report on the effect that scale would have on LCOE. No account is taken in the present work of the effect of scale on RSPOE.

Costs of electricity given in the Parsons report include capture, drying, and pressurization of CO_2 to about 1222 psia (8.43 MPa) at which point it is ready for pipeline transport. In the present analysis an additional cost of $3.00/tonne of CO_2 has been added for transport from generating station to oil field (Wallace, 2000).

Authors of the Parson report estimate that gas- or coal-based generating technologies that include CO_2 capture will be ready for commercial deployment by 2010. We concur in this assessment. Technology development is not the principal obstacle to deployment, however. Rather it is the need to develop assurances concerning safety and permanence for CO_2 storage in depleted oil fields and to build acceptance among political leaders and the population at large for this type of activity. Some additional regulatory structure may also be required.

The Parsons report shows how LCOE was calculated for each of the technologies of interest in the present study for assumed constant values of fuel prices. The analysis follows the familiar approach developed by EPRI. Components of LCOE included in the analysis are shown in Slide 8, although as noted above the Parsons study did not treat CO_2 as a byproduct. LCOE calculations assumed a book life of 20 years.

The present work treats cost of fuels (natural gas and coal) and the value of CO_2 as variables over a nominal 20 year period starting in 2010. For each year of the study period, the required selling price of electricity is calculated in order to satisfy the expected rates of return for three classes of invested capital. These three investment classes and the expected returns are shown in Slide 10. As explained below, the value of CO_2 produced was assumed to be fixed by the price of oil in each year of the study. The value of the CO_2 byproduct credit was fixed by the assumed capacity factor and the assumed price of oil. Thus, RSPOE was computed to satisfy the capital structure for power plants both with and without practice of capture and sequestration.

Of course in a deregulated electricity market the actual prices that a generator receives for electricity could be higher or lower than the RSPOE. If electricity revenue is higher than RSPOE, the financial return would be greater than that specified in the capital structure shown in Slide 10. If electricity revenue is lower than RSPOE, the financial return is lower than that shown in Slide 10; it is possible that a net loss would be realized. Rates of return on bonds and preferred stock are fixed, so all uncertainty in financial performance is borne by holders of common stock. We have computed the expected rate of return on common stock equity as follows. Rates of return on debt and preferred stock as specified in Slide 10 are treated as fixed costs. The difference between expected electricity revenue and required selling price to cover debt and preferred stock interest and dividend payments is computed. This difference, which could be positive or negative, is divided by the amount of common stock equity. The result is the expected rate of return on common stock equity.

The value of CO_2 in any year is calculated using a relation developed by Martin and Taber (1992) for practice of CO_2 EOR in the Permian Basin.

$$\text{Value of } CO_2 \text{ ,\$/Mcf} = 0.50 + 0.020 * (\text{oil price, \$/bbl}) \qquad (1)$$

Equation 1 was modified to express value in year 2000 dollars by use of the GDP price deflator.

An idea of the relative significance of revenues from electricity and sales of CO_2 can be gleaned from Slide 11.

To compute RSPOE in California in 2010 and following years it is necessary to specify expected prices for natural gas, coal, and oil. Ideally, the prices used for natural gas and coal would be authoritative predictions for delivered prices to California generators. For the period 2010-2020 we have used prices predicted using the Energy Information Agency's *AEO2002* base case version of the National Energy Modeling System (NEMS). At the time we were doing our analysis, predictions for the prices of gas and coal delivered to California generators were not available, so we used the best aggregated data available. For natural gas these were prices to utilities in the Pacific region. Since at present there is no coal used by utilities for power generation in California, we used predictions of U.S. average delivered coal prices to utilities. For oil, we used World Oil Price as defined by the EIA.

Price predictions contained in *AEO2002* extend only to the year 2020, so it was necessary to otherwise specify expected prices from 2021-2030. For natural gas, coal, and oil the base case assumption we used is that there would be no price change over this 10 year period, that is, the price predicted for 2020 would remain constant for the rest of the study period. For natural gas and oil we also created sensitivity cases in which the price trends predicted in the period 2015-2020 were assumed to continue in a linear fashion to 2030. The predicted prices with base case and sensitivity case extensions to 2030 are shown in Slides 15 and 16. Note that actual prices for all three fuels for the same geographic region, where applicable, are shown for the year 2000 on the Slides. The EIA's NEMS predicts that for all three fuels, prices through 2020 will stay below those of 2000.

To estimate rate of return on common stock equity it is necessary to predict electricity revenue per kWh received by generators as well as RSPOE. As with the prices of fuels, ideally these would be revenue received by California utilities. Again similarly to the case for fuel prices, EIA-NEMS predictions of electricity revenue to California utilities were not available. Predictions for the CNV region of NERC (National Energy Reliability Council), which includes California, were used. The predicted prices are shown on Slide 15. EIA-NEMS predicts no change in electricity revenue per kWh from 2015-2020, so there is no sensitivity case with respect to this variable. Note that similarly to fuels, EIA-NEMS predicts that electricity revenue per kWh will be below those of year 2000 throughout the period 2005-2020.

Measures of uncertainty for the prices used in the analysis were computed from historic data for the period 1990-2000. Prices for World Oil and coal delivered to U.S. utilities are shown in Slide 12. Annual average revenue per kWh received by California utilities and average delivered cost of natural gas to CA utilities are shown in Slide 13 for the same period. For each data set, standard deviation was computed in two ways. One way was to compute the linear regression coefficient for the data versus time and the attendant standard error of estimate. The other way treated the data as uncorrelated with time and simply computed the standard deviation for the 11 datum points in each data set. The smaller estimate of the standard deviation computed in the two ways was adopted. Thus, the standard error of estimate was used for prices of electricity and coal, and standard deviation based on uncorrelated data was used for oil and natural gas. These standard deviations are listed in Slide 14.

The reason for computing standard deviations in two ways is to allow for the possibility that the data follow a trend with time, in which case deviation of datum points from the trend line is smaller than differences among the data values. The correlation coefficient for coal price data in Slide 12 versus time is negative 0.994, for instance, indicating good linear correlation. The standard error of estimate expresses the standard deviation of data from the linear correlation. Its value is $0.024/million BTU as shown in Slide 14. The standard deviation of the 11 data points for coal price without regard to possible correlation is $0.200/million BTU.

The dotted lines above and below the graphs for predicted prices in Slides 15 and 16 represent one standard deviation. From the standard deviations listed in Slide 14 and from the graphs in Slides 15 and 16 it is evident that the relative uncertainty of coal price is small compared to the others. The relative uncertainties for oil and gas prices are large, that for natural gas being the largest. Uncertainty in value of CO_2 is expressed in this analysis via oil price, but from Slide 11 it is seen that CO_2 revenues are considerably less than electricity revenues for both NGCC+S and IGCC+S. Thus we anticipate that the uncertainties in RSPOE and rate of return will be greater for gas-based than coal-based generators.

Predicted RSPOE for all three technologies modeled at 65% capacity factor are shown in Slide 17, and at 80% capacity factor in Slide 18. Dotted lines above and below the expected values represent one standard deviation. For computations of NGCC without capture (labeled NGCC-H in Slides 17 and 18), only natural gas price is treated as a probabilistic variable. For computations of IGCC+S, both coal price and World Oil Price are probabilistic variables. For computations of NGCC+S, both natural gas and World Oil Price are probabilistic variables. The results were obtained by performing 500 simulations for each year in the study period. Probabilistic variables were considered to be normally distributed, with the standard deviations shown in Slide 14. Of the several probabilistic variables used in the study, only price of electricity and price of natural gas were judged to be causally related. The correlation coefficient for these two data sets shown in Slide 13 was calculated, and the appropriate correction made for performing MonteCarlo simulations for NGCC and NGCC+S systems.

Slides 17 and 18 show that the RSPOE for NGCC is the lowest for the three technologies over the entire study period, indicating this is the lowest cost generating approach of the three studied. Next highest is IGCC+S, and highest is NGCC+S. The largest standard deviation is for NGCC+S, being marginally larger than that for NGCC, and several times larger than for IGCC+S. Computed values of RSPOE for all technologies are affected most by predicted prices of their respective fuels. Thus, RSPOE for the gas-based technologies increase with time, while that for coal-based IGCC+S decreases. The sensitivity cases for both gas-based technologies show higher RSPOE than the base case in the period 2021-2030 because of the assumed higher cost of fuel. By contrast, the sensitivity case for IGCC+S is lower than the base case because of the assumed rising price of oil, which results in higher value for CO_2 .

Because NGCC+S exhibits both a larger RSPOE than either of the other two technologies and also the highest uncertainty in RSPOE, it is clearly the poorest technology from the point of view of the variables considered in this study. In a general way, NGCC+S picks up some of the disadvantages normally associated with coal-based generation technologies (i.e., relatively high capital cost) while retaining a large drawback of all gas-based technologies, namely dependency on relatively high priced fuel. Because the heat rate of NGCC+S is higher than that of NGCC, the dependence on gas price is actually intensified for NGCC+S. Revenues from CO_2 are not large enough to overcome the higher capital costs and higher heat rates incurred. The technology NGCC+S is not taken forward in the present study.

Predicted rates of return on common stock equity for NGCC and IGCC+S are shown in Slide 19 for 65% capacity factor and Slide 20 for 80% cf. For these calculations all the variables affecting RSPOE are relevant, and there is an additional probabilistic variable, the expected electricity revenue per kWh. The expected rate of return is higher for NGCC than for IGCC+S for both capacity factors over the entire study period. For the first two years of the study period, 2010 and 2011, the rate of return for NGCC is startling high, over 60% annual return at 80% capacity factor. In subsequent years the rate of return moderates due to expected reductions in revenue for electricity and increased natural gas price. The sensitivity case for NGCC exhibits a declining rate of return due to assumed increasing gas cost. Even at the lower capacity factor studied, 65%, and at the highest price of natural gas considered via the sensitivity case, the expected rate of return on common stock equity is a respectable 15%.

The most important result for the expected rate of return on common stock equity for IGCC+S is that the return is positive without any regulatory premium for avoided carbon emissions. In fact, at 80% capacity factor, its expected rate of return beats the target value of 8.74% shown in Slide 10. At 65% cf the expected rate of return comes close to the target value.

Similarly to the case for NGCC, expected rate of return on common stock equity for IGCC+S declines significantly after the initial two years in the study period, due to the expected decline in electricity revenue per kWh. Due to the expected steady decline in coal price through 2020, the rate of return for IGCC+S exhibits a small increase after 2015.

The standard deviation for expected rate of return on common stock equity is several times larger for NGCC than for IGCC+S. For none of the conditions shown in Slides 19 or 20 does the lower dotted line that represents one standard deviation go into negative territory. Recall that for normally distributed variables there is a 68% probability of a variable falling between plus/minus one standard deviation, and a 95% probability of it falling between plus/minus two standard deviations. At 65% capacity factor, both technologies exhibit negative returns at two standard deviations. Although the expected rate of return is higher for NGCC than for IGCC+S, so also is the probability of a negative return. This is a consequence of the larger standard deviation of NGCC.

DISCUSSION

As noted above, it would have been preferable to use predicted prices that are solely for the California market, but these were not available. Such prices were available for historic data, and as was mentioned they were used to compute standard deviations. How significant to the analysis was use of the CNV region of NERC for expected electricity revenue, and the Pacific region of NEMS for the expected delivered price of natural gas to utilities, instead of values for California only? Some insight into that question can be gained by comparing actual prices for electricity and natural gas for the year 2000 for California and for the "California proxies." Refer to Slides 13 and 15 and compare data for both gas and electricity for year 2000. See that the price of electricity in CA was 8.53 cents/kWh while that for the CNV region was 6.50 cents/kWh. For natural gas, the price in CA was \$5.77/million BTU, and the price in the Pacific region was \$4.95/million BTU. In both cases the prices were significantly higher for California than in the larger regions that contained California.

If expected prices for electricity in California are higher than were assumed in this study it would have the effect of making all three technologies appear more profitable. Higher natural gas cost in California than was assumed in this analysis would degrade the results for both NGCC and NGCC+S but would not affect IGCC+S.

In a similar vein, this analysis used a relation for CO_2 value to oil price that was developed for the Permian Basin. As shown above, there are multiple large sources of CO_2 available in the Permian Basin, but at present none in CA. It was noted above that CO_2 being used in an EOR pilot project in California is selling for \$3.50/Mcf, a value considerably above that calculated using Equation 1, which relates CO_2 value to oil price. Perhaps the price of CO_2 used for EOR in California will be higher than that in the Permian Basin. This would improve the predicted financial performance of IGCC+S and to a smaller extent of NGCC+S.

Concerning coal, a national average delivered price to utilities was used in this analysis. The components that determine this average have large differences, western coal being considerably cheaper than eastern coal. Coal burning utilities in California would certainly use western coal. On the other hand, transportation from the Powder River Basin, source of large reserves of western coal, to California may be higher than the national average for utilities. Additional analysis could develop a better estimate of expected price of coal in California.

REFERENCES

Bureau of Economic Geology (Texas), "Reduction of Greenhouse Gas Emissions through Underground CO2 Sequestration in Texas Oil and Gas Reservoirs," 1999.

"Evaluation of Innovative Fossil Fuel Power Plants with CO_2 Removal," EPRI, Palo Alto, CA, U.S. Department of Energy—Office of Fossil Energy, Germantown, MD and U.S. Department of Energy/NETL, Pittsburgh, PA: 2000. 1000316.

Frey, H.C. and Z. Iwanski, "Methods for Characterizing and Managing Technological Risks in Advanced Power Generation Systems: Application to Gasification Repowering in Poland," Proceedings Gasification Technology Conference, EPRI, 1997.

Martin, F.D. and J.J. Taber, "Carbon Dioxide Flooding," J. Petroleum Technology, 396-400, 1992.

Montgomery, S., M. Morea, M. Emanuele, and P. Perri, "San Joaquin basin is scene of new effort to evaluate EOR in Monterey," Oil & Gas J., 98.39, 2000.

Moritis, G., "Future of EOR & IOR," Oil & Gas J., 99.20, 68-73, 2001.

Wallace, D., "Capture and Storage of CO_2: What Needs to be Done?" COP 6 The Hague, International Energy Agency, Paris, 2000.

Prospects for Early Deployment of Power Plants Employing Carbon Capture

John Ruether[1], Robert Dahowski[2], Massood Ramezan[3], Charles Schmidt[1]

1. National Energy Technology Laboratory, USDOE
2. Pacific Northwest National Laboratory, Battelle
3. National Energy Technology Laboratory, SAIC

Electric Utilities Environmental Conference

Tucson, AZ   January 22-25, 2002

CO$_2$-EOR: The U.S. Landscape

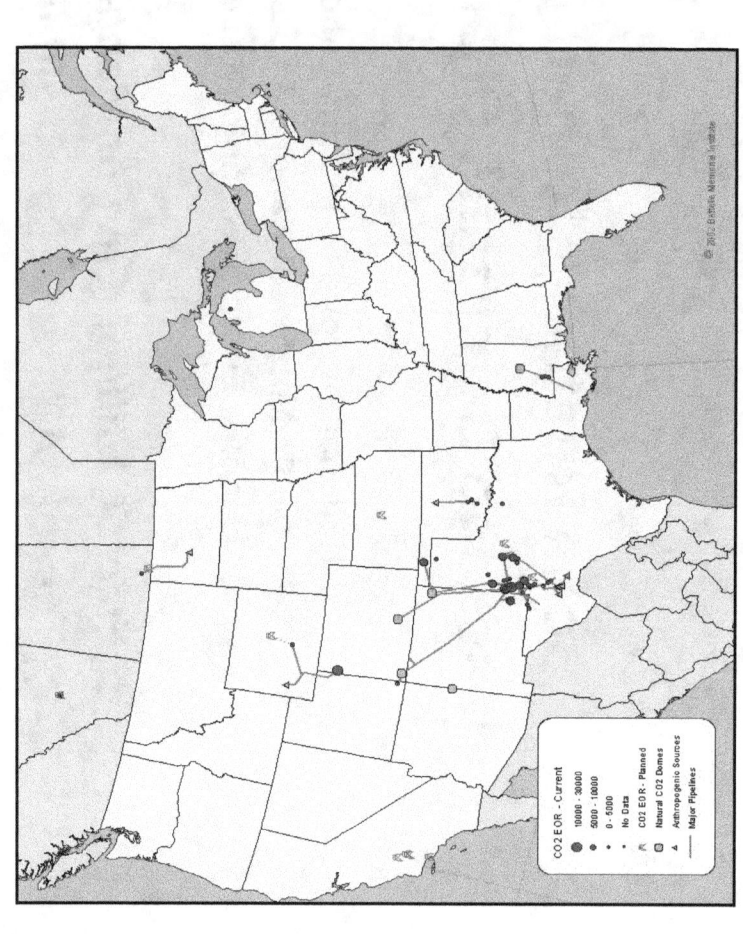

- 66 Projects: > 190,000 bbl/day enhanced production

- 5 CO$_2$ Domes: > 1300 MMcfd, 30 TCF recoverable reserves (50+ years worth)

- Other CO$_2$ Sources

- CO$_2$ Pipeline Infrastructure

CO$_2$-EOR: The Permian Basin

- 47 Projects: > 155,000 bbl/day enhanced production

- 3 Domes Supplying Majority of CO$_2$ (> 1Bcfd)

- Gas Processing Plants Supplying Remainder

- CO$_2$ Pipeline Infrastructure (1900+ Miles)

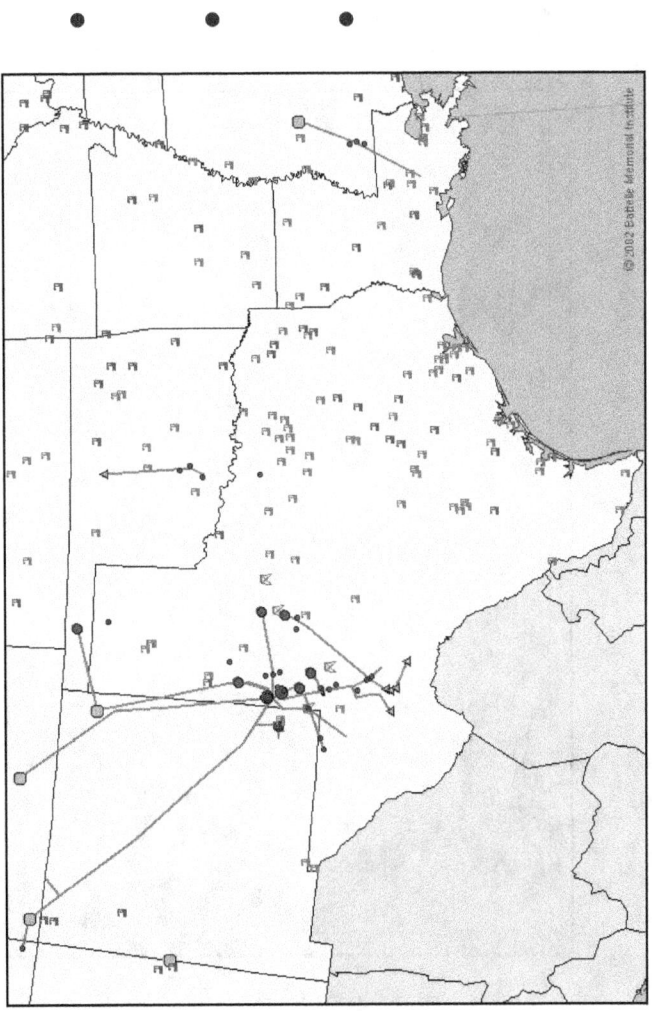

CO$_2$-EOR: California Prospects

- Many experts believe next largest opportunity outside Permian Basin
- CO$_2$ Demand Estimate: 3-5 Tcf
- Mature Fields in San Joaquin and Los Angeles Basins
- Pilot Study On-Going
- Success Could Lead to Recovery of Billions of Barrels of Trapped Oil

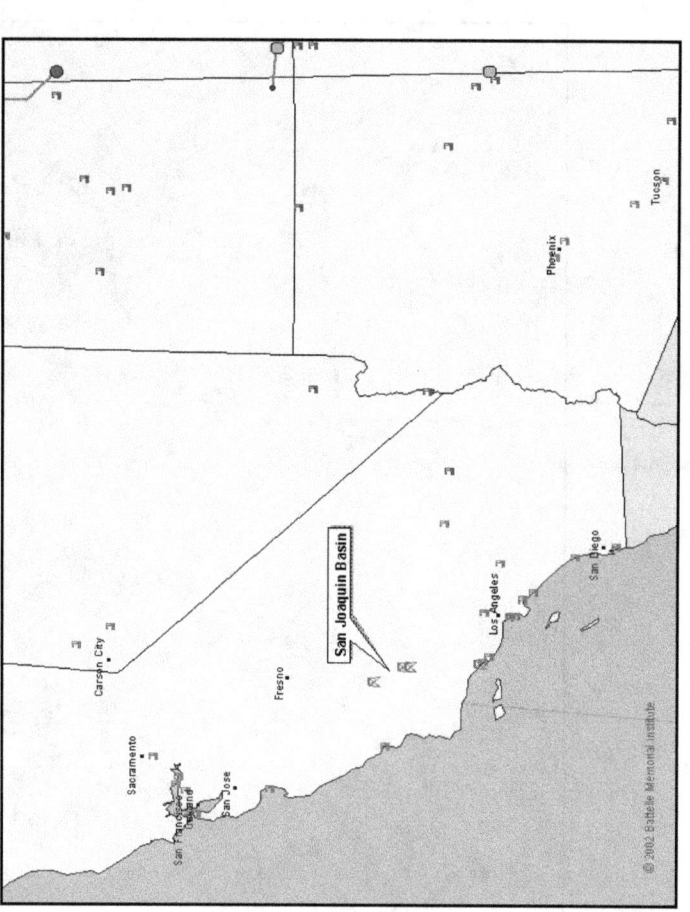

San Joaquin Basin

© 2002 Battelle Memorial Institute

Some Risks in Building and Operating Electric Power Generators

- **Technical Risk**
  - Construction cost overrun
  - Delay in start up
  - Equipment fails to achieve design performance
  - Unscheduled down time
- **Regulatory Risk**
  - Construction permits delayed/denied
  - Operating permits delayed/redefined
  - Emission standards tightened
- **Supply Risk**
  - Price increase for fuel, etc.
- **Market Risk**
  - Reduced demand/reduced selling price of products

Cost & Performance Data for Fossil Generators

Technology	Thermal Efficiency, HHV, %	Carbon Emissions, kg CO2/kWh	Total Plant Cost, $/kW	LCOE @ 80% cf, Mills/kWh
NGCC-H	53.6	0.338	496	30.7
NGCC-H 90% capture	43.3	0.04	943	48.8
IGCC-H	43.1	0.718	1263	45.1
IGCC-H 90% capture	37.0	0.073	1642	56.4

Source: "Evaluation of Fossil Fuel Power Plants with CO2 Removal," EPRI, 2000
http://www.netl.doe.gov/product/power1/gasification/30_publications.htm

Study Methodology

- **Fossil generators practicing capture & sequestration commercial in 2010.**

- **Plant book life 20 years.**

- **"AEO2002" NEMS output to estimate expected prices in California (*or a region that includes CA*)**

  - Price of electricity received by generators

  - Price of natural gas to generators

  - Price of coal to generators

  - Price of World Oil (determines value of CO_2 based on linear correlation developed for Permian basin)

Study Methodology (2)

- **Historic data for prices to estimate variability (standard deviation) in 2010-2030.**

- **Expressions for Required Selling Price of Electricity (RSPOE) for**
  - NGCC
  - NGCC+S
  - IGCC+S

- **Cost components of RSPOE**
  - Fixed O&M
  - Var. O&M
  - Consumables
  - Byproduct credit including CO_2
  - Fuel
  - Capital charges

Study Methodology (3)

- In expressions for RSPOE, fuel costs and CO_2 value are probabilistic variables.

- In expressions for Return on Common Stock Equity, RSPOE and price of electricity received by generators are probabilistic variables.

- All prices expressed as constant year 2000 dollars and cents.

Capital Structure for Plant Investment

	Percent of Total	Rate of Return	
		Current $	Constant $
Debt	45	9	5.83
Pref. Stock	10	8.5	5.34
Com. Stock	45	12	8.74
Total	100		

Typical Product Revenue per Million Btu Fuel Consumption, Dollars

(6 cent/kWh electricity, $19/tonne CO_2, or $1.00/Mcf)

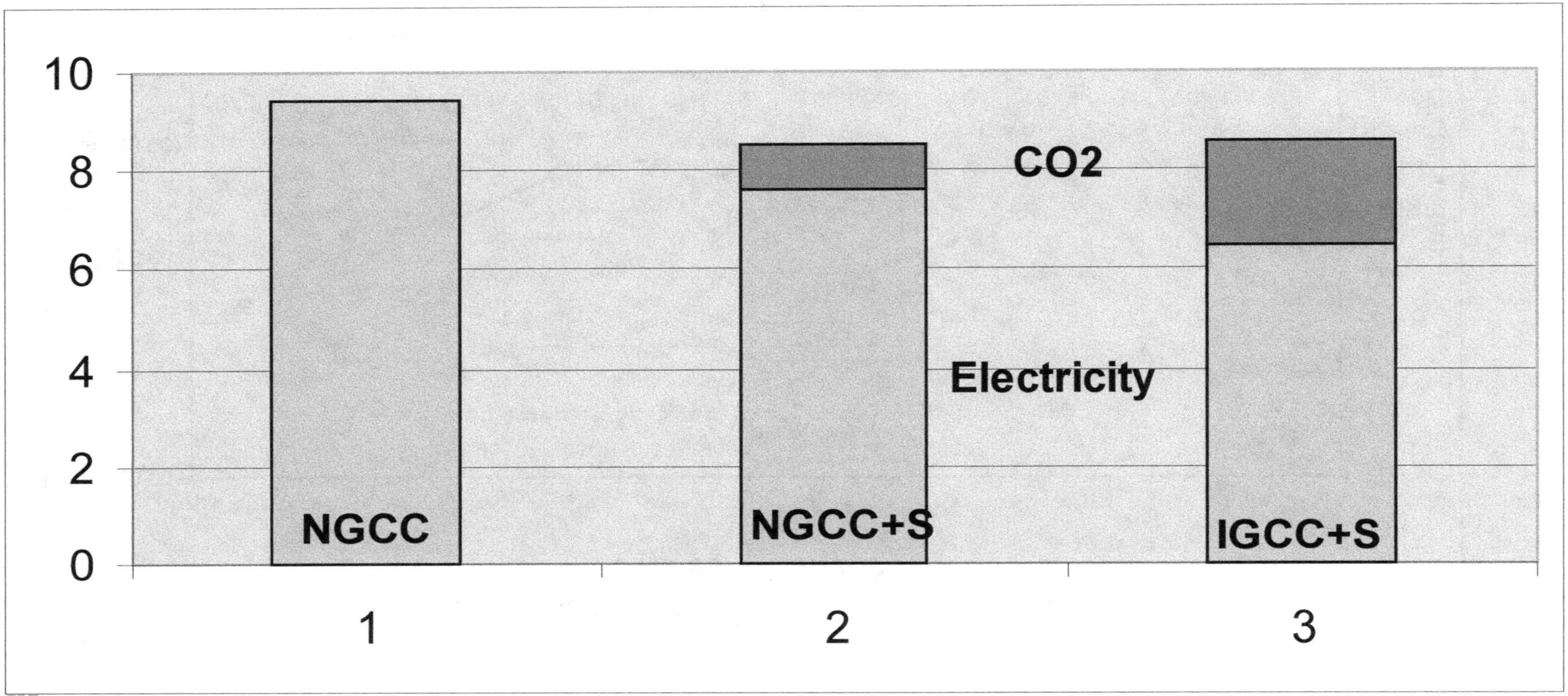

Historic Prices Used to Estimate Variability

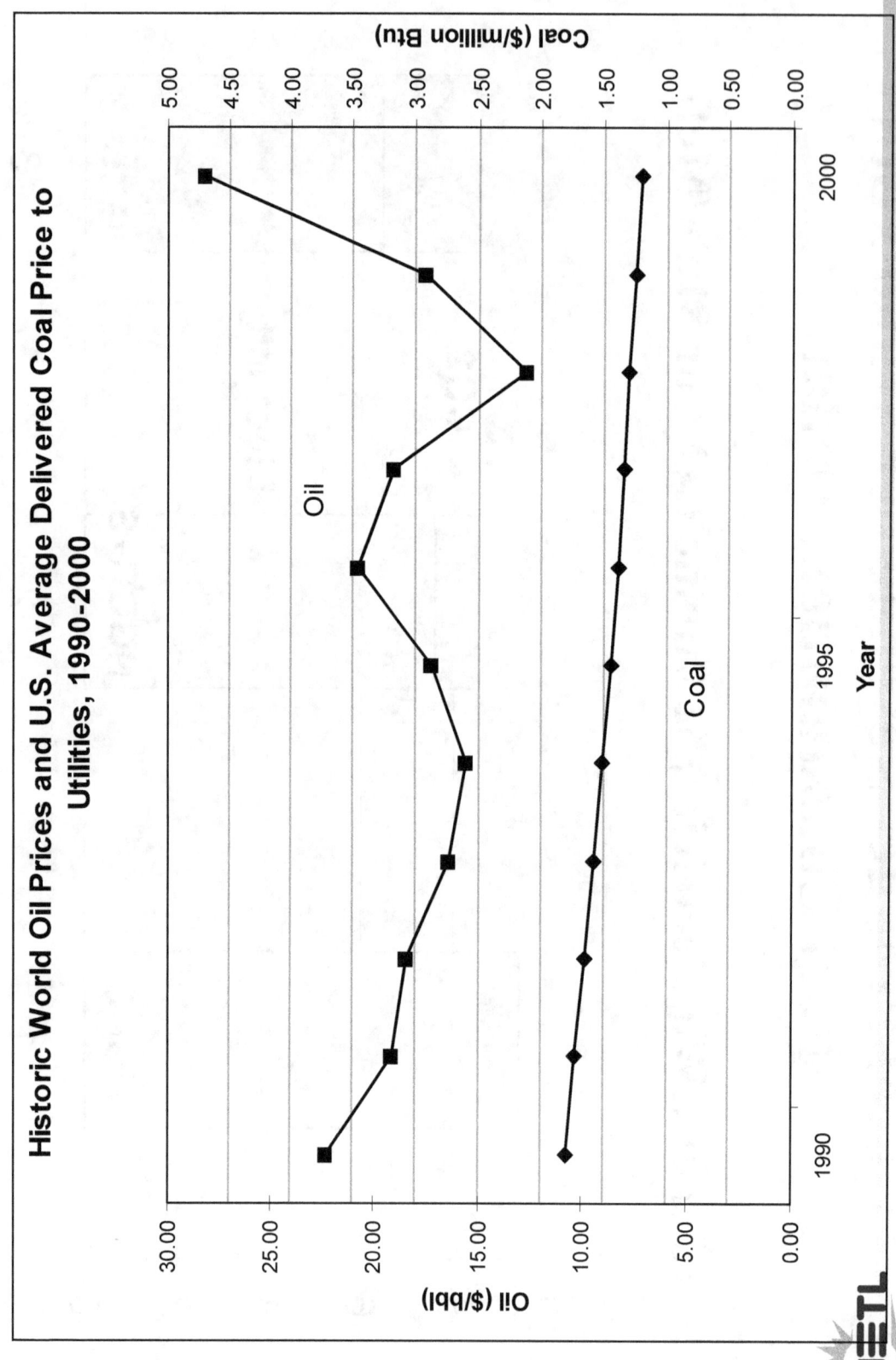

Historic World Oil Prices and U.S. Average Delivered Coal Price to Utilities, 1990-2000

Historic Prices Used to Estimate Variability

Historic Annual Electric Utility Average Revenue per kWh and Delivered Cost of Natural Gas to Utilities. California, 1990-2000

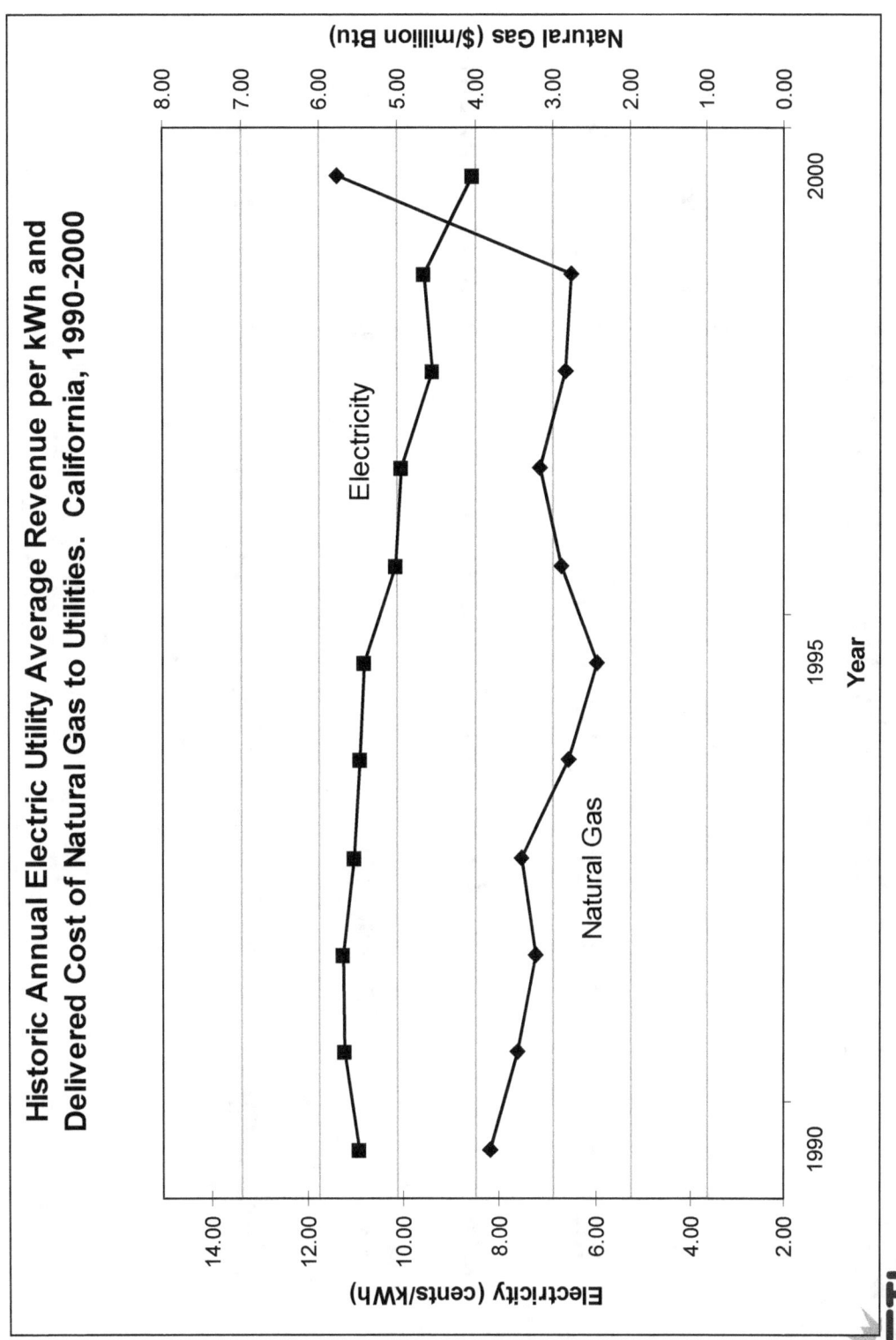

Descriptor - include initials, /org#/date

Price Variability, 1990-2000

Price	Data Set	Std. Dev.
Nat. Gas	Deliv. Cost to CA Utilities	$0.90/mill. btu
Coal	Deliv. Cost to U.S. Utilities	$0.024/mill. Btu
Oil	World Oil	$4.03/bbl
Electricity	Avg. Revenue to CA Generators	0.37 cent/kWh

Predicted Prices with Std. Devs.

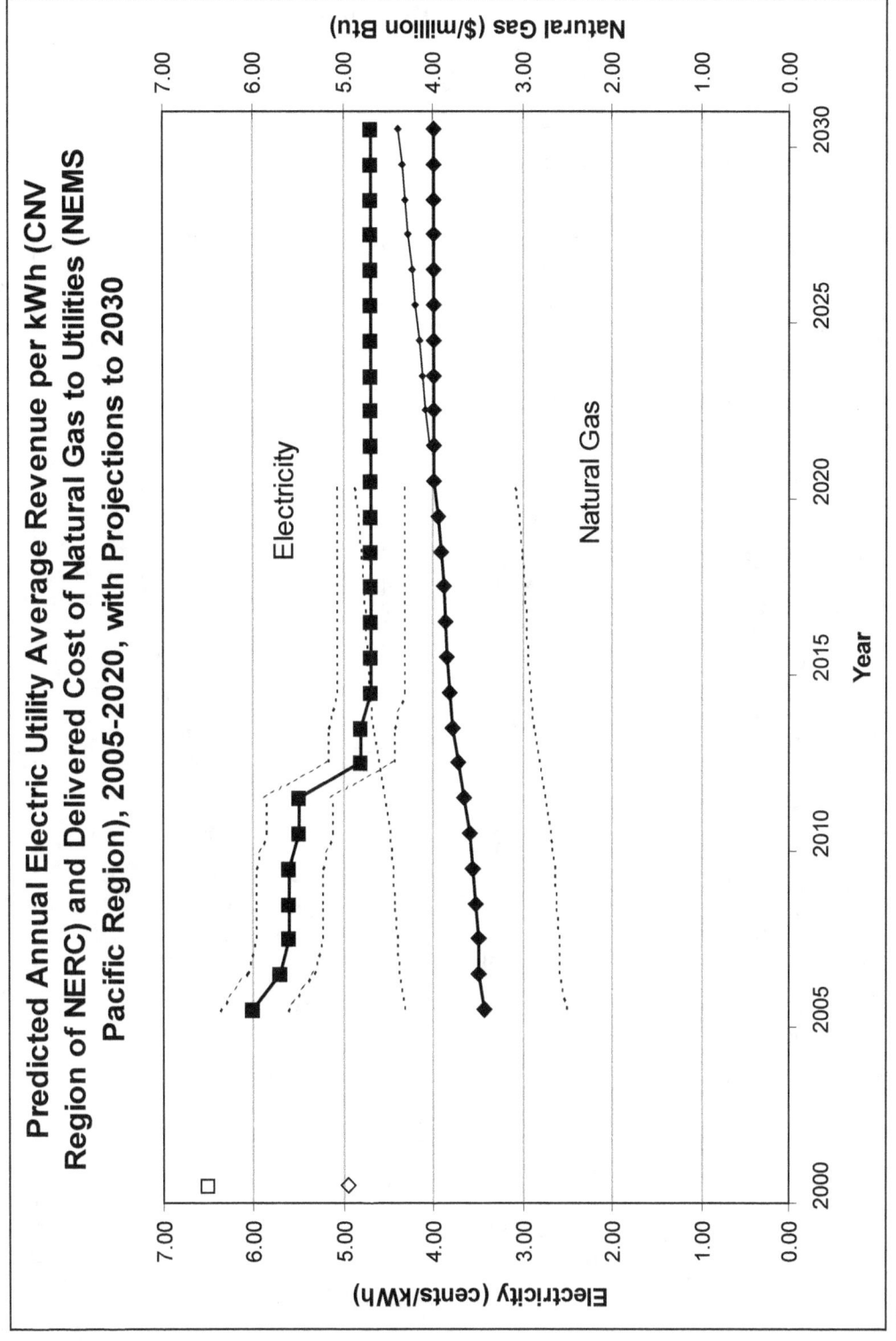

Predicted Annual Electric Utility Average Revenue per kWh (CNV Region of NERC) and Delivered Cost of Natural Gas to Utilities (NEMS Pacific Region), 2005-2020, with Projections to 2030

Predicted Prices with Std. Devs.

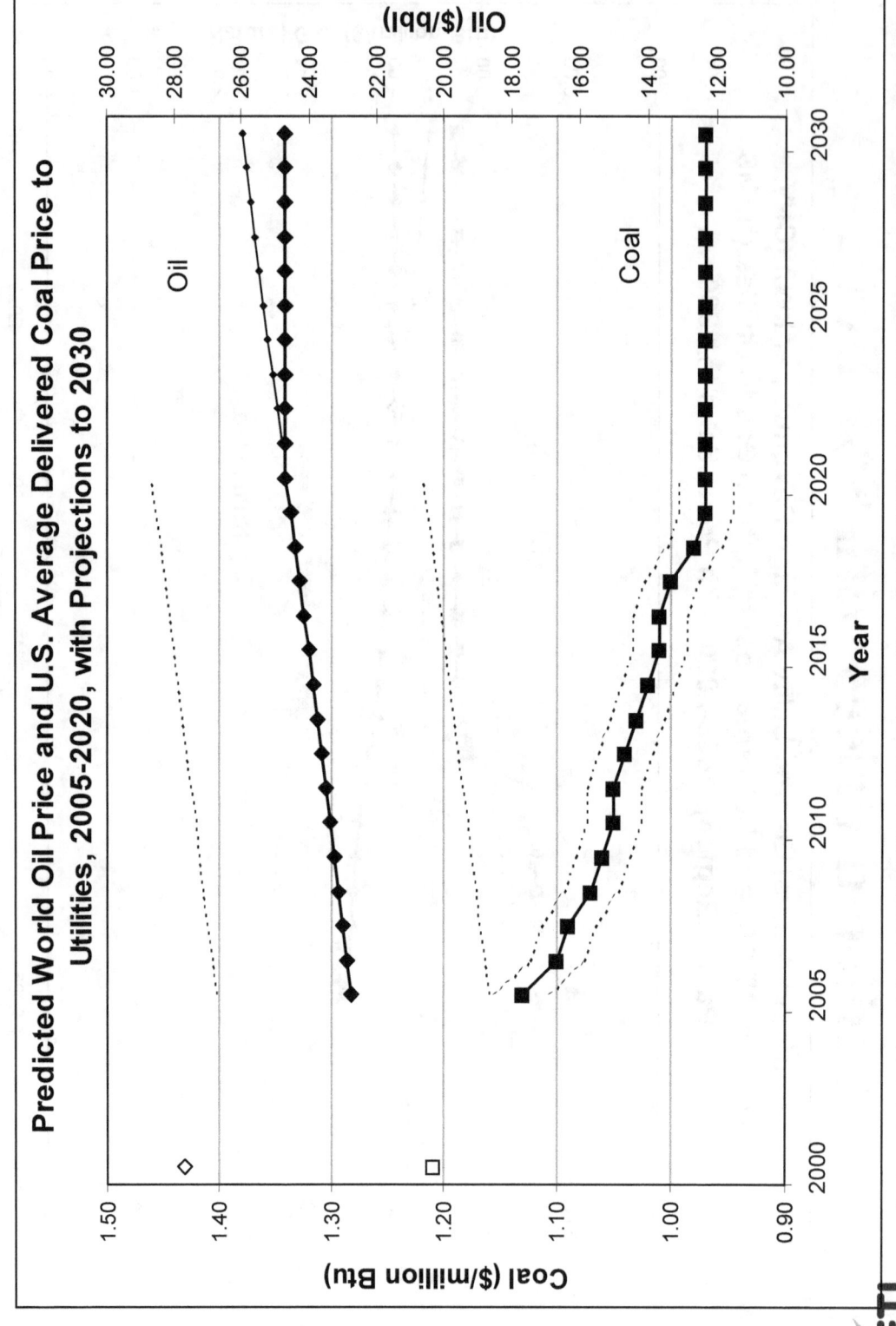

Predicted World Oil Price and U.S. Average Delivered Coal Price to Utilities, 2005-2020, with Projections to 2030

Predicted Required Selling Prices of Electricity with Std. Devs.

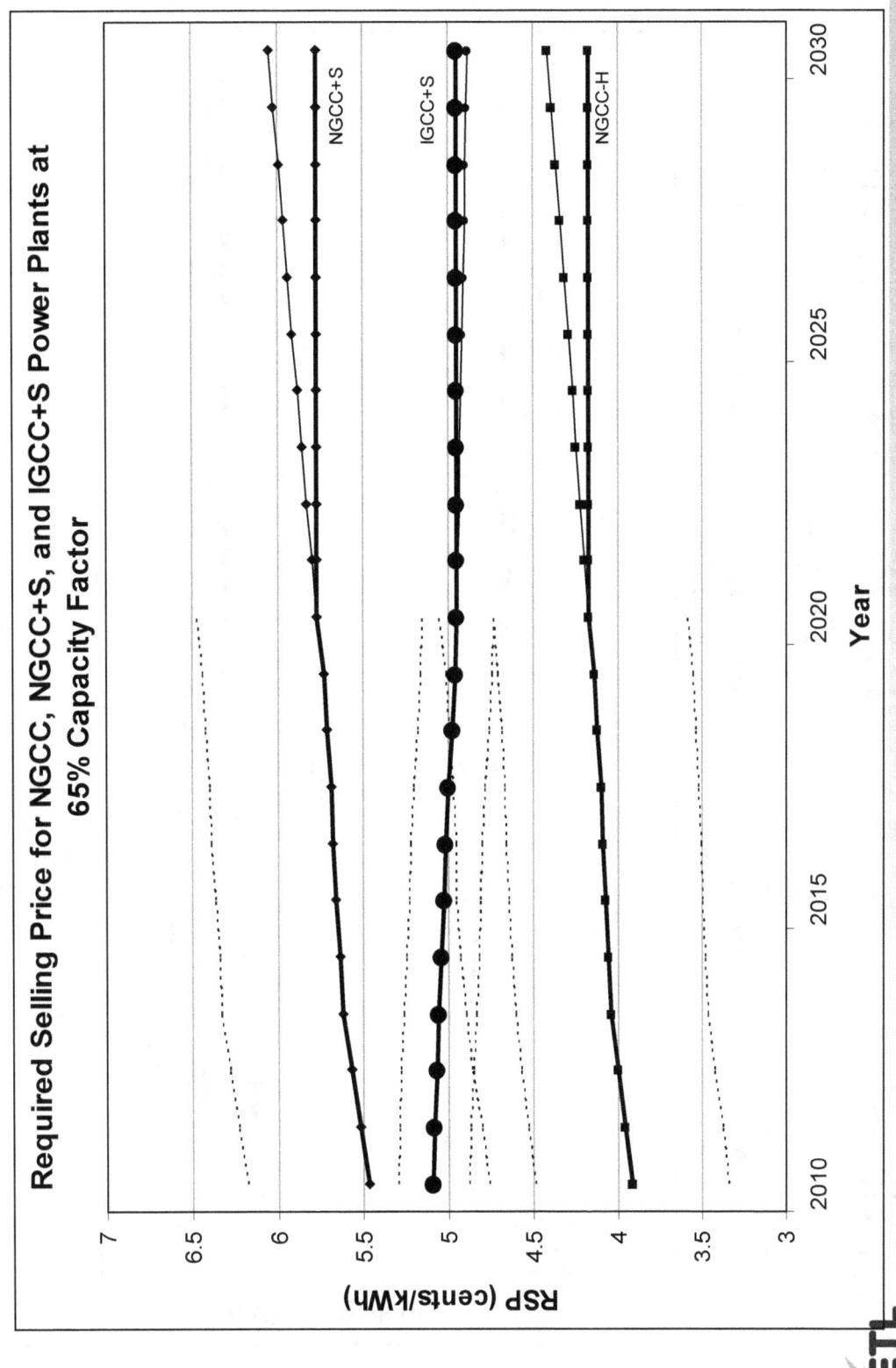

Required Selling Price for NGCC, NGCC+S, and IGCC+S Power Plants at 65% Capacity Factor

Predicted Required Selling Prices of Electricity with Std. Devs.

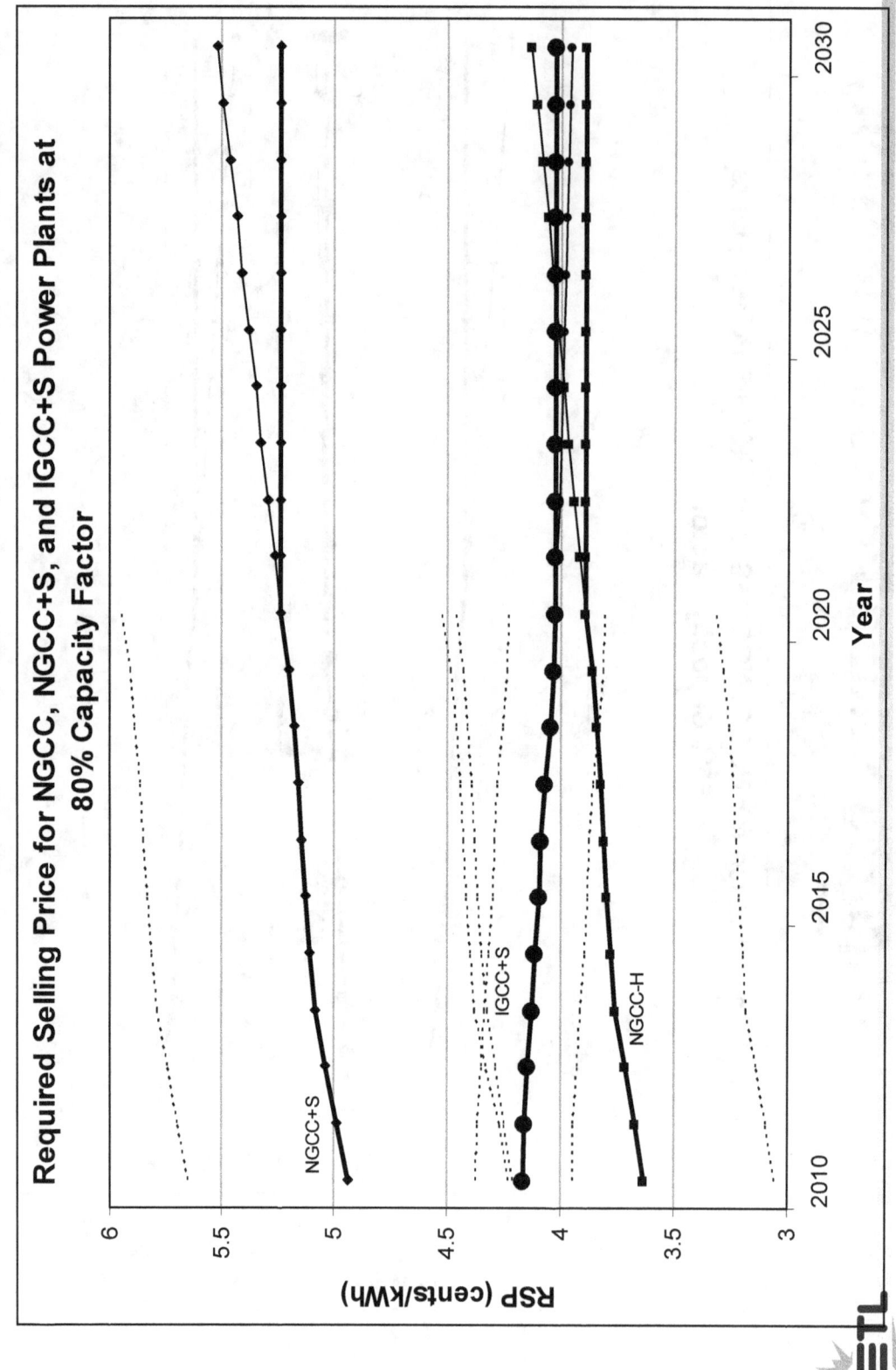

Required Selling Price for NGCC, NGCC+S, and IGCC+S Power Plants at 80% Capacity Factor

Predicted Return on Common Stock

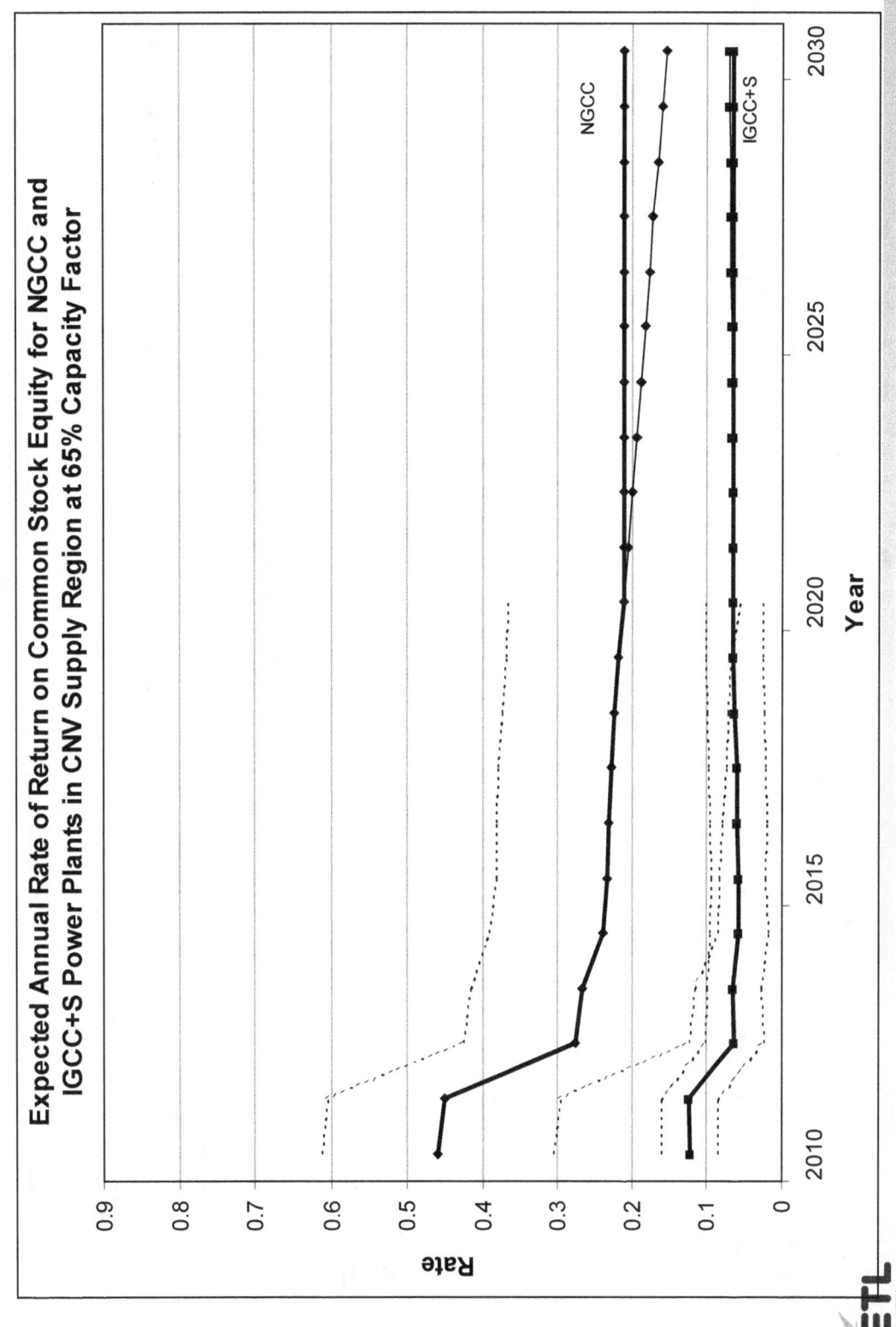

Expected Annual Rate of Return on Common Stock Equity for NGCC and IGCC+S Power Plants in CNV Supply Region at 65% Capacity Factor

Predicted Return on Common Stock

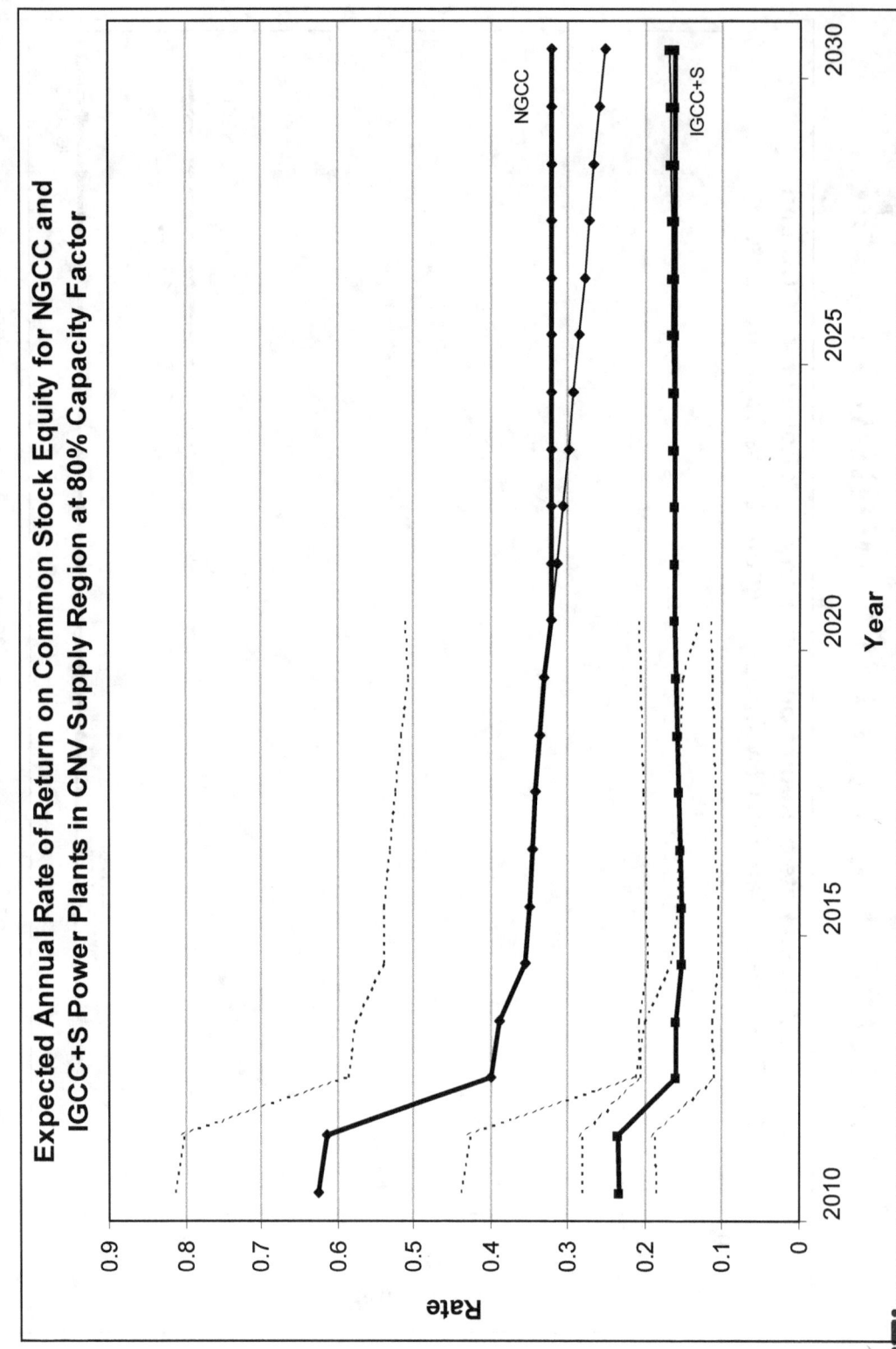

Expected Annual Rate of Return on Common Stock Equity for NGCC and IGCC+S Power Plants in CNV Supply Region at 80% Capacity Factor

Results

- When there is a market for CO_2, IGCC+S is profitable without regulatory incentive for carbon capture.

- NGCC has lowest RSPOE and highest return on investment over entire period.

- NGCC+S has highest RSPOE and greatest uncertainty in RSPOE, i.e. it is Weakest Link.

- Use of expected prices specific to CA could change results, probably in favor of IGCC+S.

- Standard deviation of RSPOE for NGCC three times larger than for IGCC+S.

- Probability of negative return on common stock greater for NGCC than for IGCC+S.